# WILDLIFE ODDITIES

# PECULIAR PLANTS

MASON CREST

**MASON CREST**

450 Parkway Drive, Suite D
Broomall, PA 19008
(866) MCP-BOOK (toll free)
www.masoncrest.com

Developed and produced by Mason Crest

First printing
9 8 7 6 5 4 3 2 1

ISBN (hardback) 978-1-4222-3526-3
ISBN (series) 978-1-4222-3523-2
ISBN (ebook) 978-1-4222-8346-2

Cataloging-in-Publication Data on file with the Library of Congress

# WILDLIFE ODDITIES

INCREDIBLE INSECTS

MYSTIFYING MAMMALS

PECULIAR PLANTS

REMARKABLE REPTILES

SHOCKING SEA CREATURES

# PICTURE CREDITS

# CONTENTS

## KEY ICONS TO LOOK FOR:

**Words to Understand:** These words with their easy-to-understand definitions will increase the reader's understanding of the text while building vocabulary skills.

**Sidebars:** This boxed material within the main text allows readers to build knowledge, gain insights, explore possibilities, and broaden their perspectives by weaving together additional information to provide realistic and holistic perspectives.

**Text-Dependent Questions:** These questions send the reader back to the text for more careful attention to the evidence presented there.

**Research Projects:** Readers are pointed toward areas of further inquiry connected to each chapter. Suggestions are provided for projects that encourage deeper research and analysis.

# PLANTS—CAN'T LIVE WITHOUT THEM

Plants live almost everywhere on Earth, from high mountaintops to the cold oceans, from dry deserts to city sidewalks. Almost every form of life depends on plants to survive. Even top predators hunt prey that eats plants. Plants also make oxygen, which we all need to breathe.

There are millions of different kinds plants all doing their part in the web of life. Whether in huge farms, vast prairies, or large forests, you can see evidence of plants— even from space!

◀◀ *A Joshua tree grows in the Mojave Desert in California, which receive less than 5 inches (13 cm) of rain a year.*

▶▶
*This flower, called Sturt's desert pea, blooms in the Australian desert after the winter rain.*

## JUST A LIGHT SNACK

We all need air, water, and food to survive. But a plant's idea of a snack comes in the form of sunlight. A plant's leaves soak up energy from the Sun's rays. Plants also get water and nutrients from the soil. Then, through photosynthesis, plants turn light and sunlight into food that helps them to grow and reproduce.

*This plant, called Rafflesia Arnoldii, has the largest flowers in the world. They grow up to 3 feet (1 m) across.*

*This strange-looking plant is called a bearded orchid. Can you see why?*

There are many different kinds of plants. Most have the same basic design, with roots, a stem, and leaves, needles, or spines. Many have flowers as well.

Unlike animals, most plants cannot move around on their own. Their roots hold them in one place. But their seeds can travel long distances, carried by the wind, by animals, or on water. Plants cannot run away from danger. So they defend themselves with weird weapons, such as prickles or poison.

# PLANTS YOU CAN'T IMAGINE

The first thing most people think about when they hear the word *plant* is green—or maybe leaves or flowers. But there is so much more to the plant world. Can you imagine a plant that looks like a pebble?

If you walked through a desert in South Africa, you would see lots of small rocks. You might think there were no plants there at all. In fact, some of the pebbles you see are plants! They look so much like stones that the animals are fooled and don't try to eat them.

In lakes, streams, and even the ocean, the green slime found in the water is also a kind of plant. It is made up of many plants called algae. Some algae live together in a big, slippery, slimy mass. Some even glow! These algae monsters grow so huge that they can be seen from space.

A very different type of algae is called a diatom. Diatoms live in water and have hard shells. Each diatom is made up of only one cell. They are so small that you could fit more than 200 into the period at the end of this sentence.

## SIDEBAR

### A WHALE OF A MEAL

Water is so much more than just blue liquid. Millions of tiny plants and creatures—some so small you can't even see them—float in the sea. They are known as plankton, and whales—the world's biggest animals—feast on them.

⬆ *This slippery slime is made up of a mixture of water, weeds, and algae. The bubbles the algae are giving off contain oxygen.*

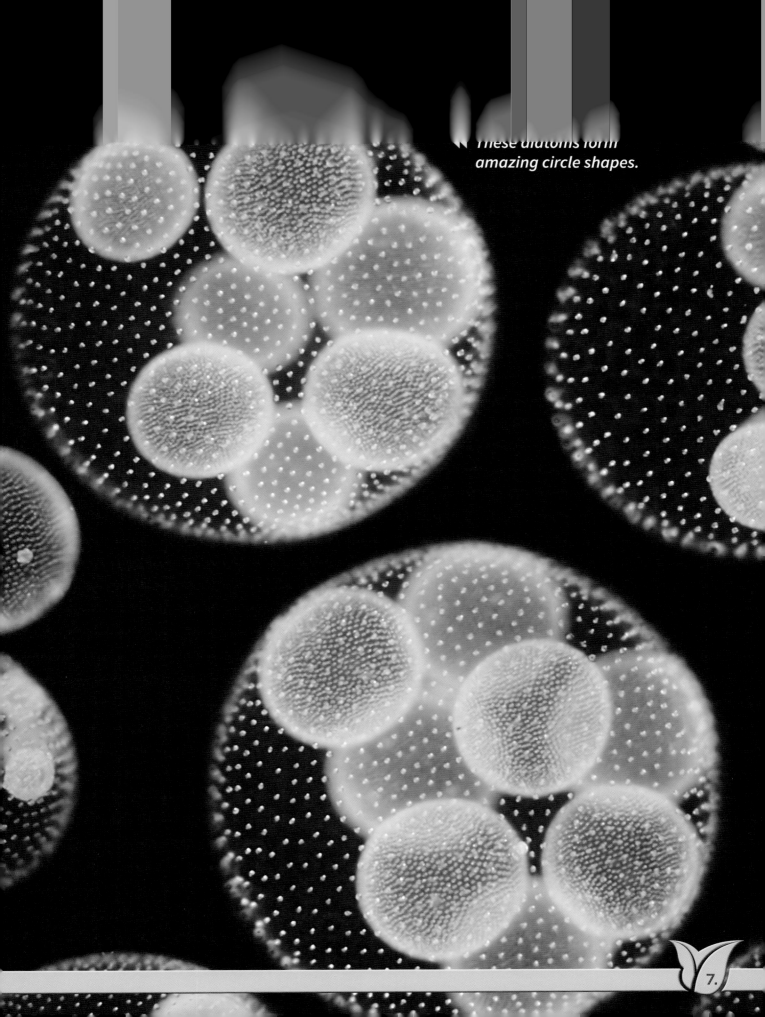

These diatoms form *amazing circle shapes.*

# SO MUCH MORE THAN PRETTY

When you put together a bouquet, you're picking one of the hardest-working parts of a plant. Most flowers have the job of making seeds so that new plants can grow. But not every flower does its job in the same way. Plants have evolved with flowers over millions of years. By now flowers come in all sorts of strange shapes and can smell like anything from perfume to rotting flesh.

When flowers get ready to reproduce, they release pollen. Tiny grains of pollen travel, carried by the wind or an insect (butterflies or bees) or another animal (birds or squirrels). If it's lucky, it will land on another flower of the same sort. Then the flowers can grow their seeds.

Flowers often look bright and smell sweet to attract insects. The insects feed on a sugary juice inside the flower called nectar. In return, insects help plants by carrying pollen along as they jump from flower to flower. One of the smelliest flowers is the Rafflesia, which stinks of rotting meat to attract flies.

Not every flower looks like roses or carnations in a flower shop or daisies in a garden. The parts of broccoli and cauliflower that you eat are also types of flowers. So are the cones on a pine tree and the sausage-shaped tops of bulrushes.

▶▶
*This is an Indonesian plant called a giant titan arum. It is made of thousands of tiny flowers. The arum flowers once every seven years.*

## SIDEBAR

### PETALS WORTH THEIR WEIGHT IN GOLD

Rose farmers in Bulgaria have to pick 1,400 flowers to make less than a teaspoonful (5 ml) of rose oil. It is an important ingredient in making perfume all around the world. Rose oil is so rare that it is more expensive than gold!

Rose Oil

*Alpine snowbells are amazing flowers. They give off heat to melt the snow so that they can grow through it.* ⏫

▶▶

*You can see grains of pollen all over this bee, which is flying from one flower to another.*

# PREDATOR PLANTS?

Most plants are happy with a little water, soil, and sunlight. But some are looking for something with a little more taste—they want meat! Most predator plants live in wet, boggy areas. They trap insects, and even small animals such as frogs, for food. You won't believe how these sneaky plants devour their prey.

The vicious Venus flytrap has deadly leaves shaped like jaws. The sharp, spiny "teeth" along the edge of the leaves aren't the real problem, though. Most of the time the plant's "jaws" stay wide open. But if an animal lands on the tiny hairs on the leaves, the trap snaps shut. The victim is trapped inside. Like other killer plants, the Venus flytrap kills its victims with a fluid that turns their bodies into juice.

▶▶

*The pitcher plant's leaves are shaped like a deep jug. Their smell attracts insects. When an insect lands on the slippery leaves, it falls into a nasty liquid in the bottom of the "jug." The liquid turns the insect's body into a juice, which the plant sucks up.*

The sundew plant has leaves covered with tiny droplets that look like dew glittering in the Sun. In fact, these droplets are a sticky trap. When an insect flies too close, its wings and legs get stuck. The leaf wraps round the insect, and the sundew starts to eat its prey.

Pitcher plants are greedy monsters. They can catch several insects at once!

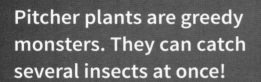

◀◀ *In this picture, a sticky sundew leaf is starting to curl itself around a trapped fly. The process takes about 30 minutes.*

▶▶ *This unlucky tree is being invaded by a mistletoe plant that sucks food and water from it. However, mistletoe doesn't usually kill its victims.*

## SIDEBAR

**TANGLE AND STRANGLE**   Make no mistake—plants are tough competitors. The strangler fig lives by wrapping itself around a tree and stealing its water and food. It also spreads out its leaves to block sunlight from its victim. It can take about 200 years, but the strangler's victim will eventually die.

# STRONG SURVIVORS

Outdoor plants have to put up with a lot—strong winds, burning sun, heavy rains, and animal teeth and toes. They have to be tough to take the abuse and survive. But some plants are not just tough—they are super tough!

A palm tree has a long, bendy trunk to help it stay alive. Palms often grow near the sea in hot, tropical countries where there are hurricanes. During these storms, winds of up to 300 miles per hour (480 km/h) smash these trees right to the ground. But many times the palm's long trunk just bends in the wind and stands up again.

▶▶

*These rope-like liana vines dangle from a tree in tropical Australia.*

◀◀
*Giant water lily pads like these can grow to more than 3 feet (1 m) across.*

Most rainforest plants grow very big because they have plenty of warmth, rain, and soil. The giant water lily, for example, lives in lakes in the Amazon jungle. Its leaves are enormous, floating pads. They are strong enough for a child to stand on.

Higher up in the rainforest, lianas dangle from the treetops. These are vines with long stems that look like ropes. Liana vines are so strong that several people can swing on them at once.

## SIDEBAR

### ROOTING FOR PLANTS

Have you ever seen a weed growing through a sidewalk or a tree growing out of a mountainside? The roots of plants can be so strong and tough that they can slowly push their way through rock and pavement.

In China, the ailanthus tree grows well in city centers even though there is much pollution. Everyone can see just how strong this tree is. Its roots can push right through cracks in concrete.

*These tough tropical palm trees in Florida are bending, but not breaking, in hurricane-force winds.* ▼

# SECRET WEAPON

When you can't move, you need to find other ways to defend yourself. Some plants use poison. When animals bite the plant, they may get sick or even die. Either way, they will not be eating that plant again!

You can tell by its name that deadly nightshade is not a plant to have for dinner. Its black berries taste sweet, but they are full of deadly poison. So are its leaves and roots.

Foxgloves are beautiful, but they make a strong poison called digitalis. Eating just a few foxglove flowers can kill you. Strangely, digitalis can also save lives. Taken in very small amounts as a medicine, it can help people with heart problems.

 SIDEBAR

**DYING TO BE PRETTY?**
Deadly nightshade is also known as belladonna, which means "beautiful lady." Italian women used to put the juice from the berries in their eyes. They thought it made them beautiful.

▲ *These pink flowers and shiny berries belong to the deadly, poisonous nightshade plant.*

These nasty-looking needles are the poison-filled spikes of a stinging nettle, seen through a microscope.

Poison ivy leaves contain a chemical that can give you a rash and make your hands swell.

Some plants can poison you even if you do not eat them. A stinging nettle has leaves covered with tiny spikes. At the bottom of each spike is a little bag of poison. When you brush against a nettle, the poison squeezes out of the bags, up the spikes, and into your skin. **OUCH!**

# FRIENDS OR ENEMIES?

Most animals and plants are not best friends. Animals eat plants, trample them, or chew them up, and use plants to build their homes. But plants have found some partners in the animal kingdom. Several insects have become quite important to plants.

The bee orchid has a flower that looks like a female bee. This disguise is to bring male bees to the orchid. When they visit, the bees pick up orchid pollen and carry it to the next flower on their travels. These bees help the orchid to grow seeds for the next generation.

*These ants live in a bull's horn acacia tree. The tree provides food, and the ants defend the tree by stinging animals that try to eat it.*

*◀◀ Some frogs spend their lives in the tiny pools that form between the branches of rainforest trees.*

Leafcutter ants, which live in North and South America, are fungus farmers. They collect leaves, then chew them up, and store them inside their nests. A fungus grows on these leaves, and the ants look after it and keep it alive. The fungus provides the ants with food.

In Africa, another kind of ant lives inside the thorns on the bull's horn acacia tree. This fierce ant bites any caterpillars or other creatures that try to eat the tree's leaves. In return, the tree feeds the ants with a kind of syrup.

*This bee orchid flower looks like a possible mate to male bees. The bees feed on the flower's nectar and help the orchid by spreading its pollen.*

 SIDEBAR

## HOME IS WHERE THE PLANTS ARE

Some people build their houses out of wood or other plant parts. And plants can provide weird homes for insects and other animals too. Some tree frogs live in the branches of rainforest trees, where rain forms little pools. Some moths spend their whole lives inside a yucca flower.

# ROOM TO GROW

Plants can't just pick up and move if it gets too crowded. And young plants can't move away when they get old enough. So plants need to spread their seeds around. Otherwise, soil, sunlight, and water will soon be in short supply. Some plant seeds travel amazing distances.

Plants such as dandelions and milkweed have seeds with fluffy "parachutes." When you, or the wind, blows on them, they fly away. Other plants hide their seeds inside delicious fruit. After birds, bats, and monkeys (and sometimes even people) eat the fruit, they travel with the seeds in their stomachs. After their journey, the seeds fall to the ground in the animals' droppings.

*A tumbleweed, or "wind witch," rolls along a road in the American West.*

## SIDEBAR

**BLOWN AWAY** One great traveler is tumbleweed, also called "wind witch." When its seeds are ripe, the tumbleweed breaks free from the soil, and the wind blows it away. Each plant scatters up to 250,000 seeds as it rolls along.

Some seeds have even weirder ways to travel. The squirting cucumber has fruits full of seeds and juice. The fruits sit in the sun getting ripe. Then, like an overfull water balloon, they suddenly burst. The juice squirts the seeds up to 18 feet (6 m) away.

Other plants are long-distance travelers. Coconut palms grow by the sea, and their fruits often roll into the water. Coconuts can drift thousands of miles (kilometers) on the waves before they wash up on a distant shore.

▶▶

*A dandelion seed floats along under its fluffy parachute.*

*This coconut has already started to sprout into a new plant while floating along in the sea and then washing ashore.* ▼

# TOTALLY TREES

Trees are the biggest plants of all—some are incredibly tall, and old tree trunks can be so wide that you could drive a car through them. Trees also live longer than most other plants. Scientists believe some have lived for thousands of years!

Baobab trees have the fattest trunks of all. They can be up to 90 foot (30 m) around. Baobabs live in dry places and use their fat trunks to store water. Very old baobabs sometimes have hollow trunks with a space inside as big as a large room.

The banyan tree has many thin trunks. As its branches spread, they put down roots that reach into the soil to form new trunks. A famous banyan tree in India has hundreds of trunks. It has spread so wide that 2,000 people can stand beneath it.

*Banyan trees sport many trunks at one time.*

*Baobab trunks contain so much water that they are squishy to touch.*

**HIDDEN TREASURES**   Believe it or not, jewels can sometimes come from trees. It has been said that about one in a million coconuts contains a coconut pearl, which is a beautiful, round, white bead.

The biggest trees of all are the redwood and giant sequoias, which grow in the United States. The tallest redwood is more than 335 feet (111 m) high, which is taller than the Statue of Liberty. The biggest sequoia has a trunk that is 15 times heavier than a blue whale—the largest animal on Earth!

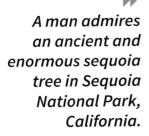

*A man admires an ancient and enormous sequoia tree in Sequoia National Park, California.*

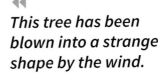

*This tree has been blown into a strange shape by the wind.*

# FUNGUS AMONG US

Fungi were once considered part of the plant kingdom. Mushrooms, toadstools, and mold truly have weird lifestyles, sharing characteristics of both plants and animals. Many grow from the ground but have no leaves, flowers, fruit, or stems. Others grow so fast that they seem to appear by magic! Many fungi are poisonous.

When you see a toadstool, you are only seeing a small part of the fungus. Beneath the soil is a huge tangle of tiny threads. The toadstool that pops up above the soil is like a flower. It releases a simple kind of seed called a spore.

◀◀
*Do trees have ears?*
*This one has an ear fungus.*

🔺 *Lichens come in a rainbow of amazing colors. A lichen is made up of a layer of fungus and a layer of algae.*

Instead of stems, leaves, and branches, some fungi have strange, round shapes. Ear fungus grows on rotting wood and looks just like a wrinkly, old ear when it is dried. Giant puffballs look like white footballs. You can probably guess what cage fungus, orange-peel fungus, and hoof fungus look like!

The mold that grows on old bread is a fungus. Another sort of fungus can grow between your toes and cause the disease known as athlete's foot. Lichens, which grow on rocks and tree trunks, are made of fungi and algae living together.

*These white truffles may not look very tasty, but some people pay top dollar to eat them.* ▼

# GREEN MAGIC?

Throughout history, and even today, people have believed that plants have magical powers—bringing health, happiness, and luck. Today, science has proved that some plants can be used as medicines. Other uses for plants are still being discovered.

People believe many strange things about plants. Hazel twigs are supposed to be good for finding water. People called water diviners use them to search for underground springs. When hazel twigs are held above water, they are said to twitch.

*Nettles and dock leaves often grow close together, just like here.*

## SIDEBAR

**HURT OR HELP?** If you are stung by a nettle, look for a large green dock leaf growing nearby, and rub it on the sting. The juice of the dock leaf makes the pain of nettle stings go away.

Mandrake is sort of nightshade. It has a root that can be shaped a bit like a person. That may be why people have strange beliefs about it. Long ago, people used mandrake to make love potions. They believed that the mandrake screamed when you pulled it out of the ground!

For hundreds of years, people have used willow bark as a painkiller like aspirin. Today, oil from the leaves of the eucalyptus tree is used to soothe sore throats. Tea tree oil helps to heal cuts, and the smell of flowers from the lavender plant helps some people sleep.

⏫ *Mandrake roots really can look like scary people. That's probably why people made up stories about them.*

◀◀

*Lavender is popular in scented waters and sachets.*

# PECULIAR PLANT FACTS

Here are some amazing facts from the weird world of plants.

## Towering tree

The tallest plant in the world is the redwood tree. The biggest redwood alive is 365 feet (111.8 m) high.

## Tiny algae and plants

Some of the smallest plants of all are single-celled algae. The smallest flowering plant, however, is *Wolffia globosa*, which floats on ponds. Each one is less than 0.04 inches (1 mm) across.

## Number one seed

The coco-de-mer palm tree has the biggest seeds in the world. They are the size of beach balls and weigh up to 44 pounds (20 kg) each. Despite being so heavy, they float when they fall into the sea.

## Chief leaf

The raffia palm has the longest leaves. They are up to 60 feet (20 m) long and are thin and feathery. The biggest flat leaves belong to the Amazonian palm and are nearly 24 feet (8 m) long. People use them as sails for small boats.

## Mammoth trunks

Some baobab trees have trunks that are 90 feet (30 m) around. Redwoods, giant sequoias, and Montezuma bald cypresses also can have trunks that are nearly as thick.

## Hair-raising flower

The rafflesia flower grows up to 3 feet (1 m) wide and can weigh 24 pounds (11 kg). Its giant, smelly petals are 0.8 inches (2 cm) thick.

## Ancient roots

The oldest tree is a bristle-cone pine in the United States, which is more than 4,700 years old. A lichen has been discovered that has been alive for 10,000 years. Scientists also have grown seeds that had been lying in frozen soil for nearly 15,000 years.

GENERAL SHERMAN

**Rainforest**
A type of forest that grows in hot, rainy regions.

**Ripe**
Fully grown, or ready to be picked and eaten.

**Species**
A group or type of plant or animal, a special name given by scientists.

**Spores**
Tiny, powdery specks that a fungus makes instead of seeds.

**Tropical**
From the tropics, which are hot, stormy parts of the world.

**Vines**
Plants with long, thin stems that either grow up things or hang down from them.

**Water diviner**
Someone who tries to find water under the ground.

 TEXT-DEPENDENT QUESTIONS

1. **What process do plants use to make sunlight into food?**
2. **What kind of tiny ocean plant helps feed whales, some of the largest animals on Earth?**
3. **What do flowers have to share before they can make seeds?**
4. **What looks a little like a plant, and acts a little like a plant, but is not a plant at all?**
5. **Where do the tallest trees in the world grow?**

▶▶

*Saguaro cactus are the largest species of cactus in the United States. This one is in an Arizona desert.*

◀◀

*These cup-shaped mushrooms are a type of fungus.*

# PECULIAR PLANT PROJECTS

**The best way to find out more about weird plants is to go and see some in a public garden, or you can even grow a weird plant at home.**

## VISIT A GARDEN

Many towns and cities have botanical gardens. They are like zoos for plants. You can visit and look at amazing plants from all over the world. Caretakers carefully look after the plants and make the right conditions to keep them healthy. There may be a tropical plant house, with palm trees and giant water lilies, or a desert house with cactuses. You might be able to see Venus flytraps, pebble plants, or rainforest flowers. You can find your nearest botanical gardens by looking online or in your telephone directory or asking at a tourist information office.

## DYE A PLANT

To see how plants suck water up their stems, try this experiment. Take a white flower such as a carnation, or a stick of celery with leaves on it, and put it in water that contains a few drops of red food coloring. How long does it take your plant to change color?

## KEEP A WEIRD PLANT

If you would like to keep a weird plant at home or in your classroom, try a Venus flytrap. You can buy them at garden centers. Ask for one that eats insects as not all of them do. Put the plant in a warm, sunny place inside a glass tank or a fishbowl with a lid. Keep its soil damp with rainwater (not tap water as Venus flytraps do not like it). Do not give it any plant food. Instead, if you can find them, put a few live ants, without big pincers, into the tank. If they do not get eaten, try feeding the plant with a tiny bit of raw meat on a pin. The plant will grow new traps as it gets bigger. Do not feed your Venus flytrap too often. They get most of the food they need from the Sun. Each trap only works about six times, and each meal lasts the plant several weeks. Keep a diary of how your Venus flytrap grows and behaves.

# PECULIAR PLANTS ON THE WEB

**Botanical Record Breakers**
**http://waynesword.palomar.edu/ww0601.htm**
Record-breaking plant facts.

**Fun Plant Activities**
**http://lifestyle.howstuffworks.com/crafts/**
**nature-crafts/plant-activities-for-kids.htm**
Fun ways to get a green thumb. Grow your own
popcorn, or plan an indoor garden.

**Grow Your Own Predator Plants**
**http://www.carnivorous--plants.com/care-of-**
**carnivorous-plants.html**
How to care for Venus flytraps and other plants that
eat meat.

**Interactive Plant Games**
**http://www.sciencekids.co.nz/plants.html**
Have fun while learning interesting stuff about plants
from around the world.

⬆ *This fly is about to meet its fate between the jaws of a hungry **Venus flytrap**.*

You can use the Internet to explore the weird world of plants. Remember that Web sites
can change, so if you cannot find all the sites shown here, do not worry. You can find
more Web sites about weird plants by typing in key words such as "plants" or "trees."

## FURTHER READING

Aaseng, Nathan. *Weird Meat-eating Plants.* Enslow, 2011.
Ganei, Anita. *Peculiar Plants.* Chicago: Raintree, 2013.
Kenah, Katherian. *Weird and Wacky Plants.* Brighter Child, 2004.
Mattern, Joanne. *Weird and Wonderful Plants.* Macmillian/McGraw-Hill, 2009.

# INDEX

In this index, bold italic font indicates a photo or illustration.